In the Fog

BY KIM THOMPSON

A Little Honey Book

Tips for Teachers and Caregivers

This book supports early readers as they decode words to learn facts and gain knowledge about the world.

Before reading, make sure students understand the sound-spelling correspondences shown below as well as the high-frequency words shown on the next page. Introduce the vocabulary words.

During reading, provide feedback and encouragement as students sound out decodable words by blending individual sounds.

After reading, talk about and write about the topic. Share the information on page 16 to help students learn more.

Letters and Sounds

New:

Sound	Spelling
/f/	f
/g/	g
short o	o

Review:

Sound	Spelling
short a	a
/k/	c
/d/	d
/h/	h
short i	i
/m/	m
/n/	n
/p/	p
/s/	s
/t/	t

Decodable Words

can, cannot, dim, fans, fits, fog, gaps, gas, got, hot, in, is, it, on, sag

High-Frequency Words

New: be, make, more, no, people, there, to, water

Review: a, help, look, out, the, this

Vocabulary Words

accidents

clouds

ground

lifts

traffic light

Water can be a gas.

This gas can make **clouds**.

ground

Clouds can sag to the **ground**.

This is fog.

Fog fits in gaps.

Fog fans out on the water.

Fog makes the **traffic light** dim.

People cannot look out.

There can be **accidents** in the fog.

People help.

lifts

It got hot.

The fog **lifts**.

No more ground clouds!

Build Background Knowledge

Fog often forms overnight when cool air condenses water vapor around bits of dust in the air. In the morning, as the Sun warms the air, the fog evaporates. When heavy fog is in the forecast, meteorologists warn people to be careful. Fog can limit drivers' ability to see. Accidents can happen. School buses cannot run safely. People can change their plans to avoid traveling in severe fog.

Written by: Kim Thompson
Designed by: Rhea Magaro
Series Development: James Earley
Educational Consultant: Marie Lemke, M.Ed.

Photographs: All images from Shutterstock

Crabtree Publishing

crabtreebooks.com 800-387-7650

Printed in China/012024/FE20231222

Published in Canada
Crabtree Publishing
616 Welland Ave.
St. Catharines, Ontario
L2M 5V6

Published in the United States
Crabtree Publishing
347 Fifth Ave
Suite 1402-145
New York, NY 10016

Library and Archives Canada Cataloguing in Publication
Available at Library and Archives Canada

Library of Congress Cataloging-in-Publication Data
Available at the Library of Congress

Hardcover: 978-1-0398-4429-2
Paperback: 978-1-0398-4511-4
Ebook (pdf): 978-1-0398-4588-6
Epub: 978-1-0398-4658-6
Read-Along: 978-1-0398-4728-6
Audio: 978-1-0398-4798-9